商业空间与住宅设计

商业空间与住宅设计

(美)马克·詹森 哈尔·戈德斯坦 史蒂文·斯库罗 编

于丽红 译

广西师范大学出版社
·桂林·

images Publishing

图书在版编目(CIP)数据

商业空间与住宅设计/(美)戈德斯坦,(美)斯库罗等 编;于丽红 译.—桂林:广西师范大学出版社,2016.1
ISBN 978-7-5495-7378-3

Ⅰ.①商… Ⅱ.①戈… ②斯… ③于… Ⅲ.①商业-服务建筑-室内装饰设计 ②住宅-室内装饰设计 Ⅳ.①TU247②TU241

中国版本图书馆 CIP 数据核字(2015)第 260602 号

出 品 人:刘广汉
责任编辑:肖 莉 于丽红
版式设计:吴 迪

广西师范大学出版社出版发行
(广西桂林市中华路22号   邮政编码:541001)
(网址:http://www.bbtpress.com)
出版人:何林夏
全国新华书店经销
销售热线:021-31260822-882/883
恒美印务(广州)有限公司印刷
(广州市南沙区环市大道南路334号   邮政编码:511458)
开本:787mm×1 092mm   1/12
印张:$21\frac{1}{3}$ 字数:30千字
2016年1月第1版   2016年1月第1次印刷
定价:268.00元

如发现印装质量问题,影响阅读,请与印刷单位联系调换。

目录

7 前言

商业空间作品

12 安达仕西好莱坞酒店
26 阿玛尼家居
30 卡尔文·克莱因公司店
36 卡尔文·克莱因内衣店
42 现代海悦酒店
44 科蒂集团行政办公室
46 DB现代酒馆
52 犀牛企业总部和展览室
54 爱姆普里奥·阿玛尼苏活区店
58 熨斗区豪华阁楼
62 霍尔特伦弗鲁旗舰店
76 霍尔特伦弗鲁旗舰店——新一代
78 霍尔特伦弗鲁男士精品店
86 卡尔捷迪克斯区hr2品牌店
90 Intermix品牌店
98 Intermix肉类加工区品牌店
104 KBond男装精品店
106 科特 / 安萨克斯店
112 路易斯登集团办公室
116 纽约奥林匹克自行车馆
120 蒙特利尔奥美集团
124 公园大道广场公共中庭
128 Rocket Dog 集团总部
136 萨克斯第五大道精品百货店
140 菲拉格慕旗舰店
146 浪花棕榈滩零售店
150 时代广场改造性再利用
154 时代广场商业开发
158 苏活区TSE品牌店
164 莱克星顿大街W酒店
168 五星级酒店与住宅
174 华盛顿新罕布什尔大道1200号
180 美洲大道1345号

住宅作品

186 贝德福德旅馆
192 中央公园西大道住宅
198 科特兰庄园住宅
200 东汉普顿小镇度假屋
212 格林威治村阁楼
220 哈德逊旅馆
222 伦敦阳台阁楼
232 曼哈顿连体别墅
240 中世纪住宅
248 红钩住宅
253 致谢
255 图片版权

前言

安德鲁·塞萨

走进詹森·戈德斯坦事务所设计的建筑——无论是住宅、零售空间、酒店或者是公共环境区域——人们都会立刻因建筑师的独特设计而感到震惊。优雅宽敞的布置吸引大量林荫道上探究的目光。充足的自然光通过精心安置的玻璃墙、窗户和天窗透入,有选择性的照亮一些元素,而使其他元素投入到阴影中,这种借来的形式适用于整体结构。不考虑项目的独特风格,过度优雅的盛行已经被丰富多样的材料和紧密关注的成品与细节所减缓。每个空间移动、每个家具、每个设备和装置背后,都存在一个精确而清晰的愿景和目的。詹森·戈德斯坦的建筑都流露出一种明确的特征:地域的独特感知。事务所给予亲密的个人空间和宏伟的公共项目同等的注意力,从整体上使建筑风格既符合个人的需求,也融入到团体之中。

例如,在传说中有名的纽约长岛东区的东汉普顿小镇中建一个带6间卧室共900平方米的度假屋。不但需要将建筑结构巧妙的融合进小环境的历史特色,也需要将传统特色与现代特色相结合。詹森·戈德斯坦为度假屋发展了一种充满美感的风格——沙克尔风格。建筑师借鉴沙克尔有力而简洁的几何图形,创造出一种低矮的三角形屋顶和灰色木瓦立面。立面的窗户都以网格形式出现,上面六块玻璃,下面也是六块玻璃。大空间中的翅形地面,脱色的橡木地板,裸露的梁柱,从视觉上将厨房、餐厅、休息区连成一个整体。大量的窗户和玻璃门允许空气和自然光进入,13米长的天窗穿过整个三角形天花板。在楼上,卧室坐落在单一长廊的一边,使光和空气的移动最大化。无论是室内还是室外,这个度假屋都将设计师对住宅的深入理解,转化成了整洁、清晰、优雅的外观和空间,显得既古典又现代。

在1995年,詹森·戈德斯坦事务所创立之初,纽约公司合伙人的实践以两大原则为基础:建筑和室内设计的完美融合;见证规模、项目和建筑类型的广阔多样性的数据库的建立和培养。

自从那时起的20年间,事务所的三个合伙人——哈尔·戈德斯坦、马克·詹森和史蒂文·斯库罗始终秉持着这些原则,从外向内和从内向外同时创造空间。这些空间大小不同、区域不同、风格不同、种类不同:既有在纽约北部的120平方米的小旅馆,也有在温哥华市中心闪耀的12,000平方米的奢侈品百货店;既有在华盛顿哥伦比亚区的复式带玻璃展区的办公楼,也有在西好莱坞带240间客房的酒店和餐厅。

不考虑这种多样化，合伙人的独特理念和工作流程始终不变，首先都是将首要的工作原则作为指导。需要进行说明的是，这一观点并不是一种美学——属于一种独特风格或学派——而是一种哲学。他们相信严格而带有强烈目的性的场所建造可以毁灭一个项目独特而唯一的选址、环境和客户的需求。

所以他们的流程证明每个部分都具有探索性、包容性和协作性。建筑设计师和客户会谈决定人们如何去体验一个地方，如何作用于一个地方，如何感觉一个地方和空间。设计师根据对话，去倾听和解释，探索和加固方案。

在流程之外，建筑设计师为每个建筑开发了一种特色。通过考虑最高比例，他们制定了规划和立面图，表现了精准度和限制。通过无数创新的方法，包括依靠自然光的力量、对周围的开放性、和透过玻璃进入蓝天而给空间带来生命力。

比如，在温哥华，建筑师和该城市的南森·艾伦玻璃工作室一起创造出独特的，看起来中间絮有软物的"枕头式"玻璃，覆盖在奢侈品零售商霍尔特·伦弗鲁旗舰店的整个立面上；在曼哈顿中心的公园大道广场，他们和德国的肖特集团合作，设计了照明玻璃管，排列在大堂的墙上。

尽管事务所已经为一些世界上最著名的时尚、零售、酒店品牌工作过，包括乔治·阿玛尼、萨瓦托·菲拉格慕、卡尔文·克莱因、和路易斯登，霍尔特·伦弗鲁和萨克斯第五大道精品百货店，以及凯悦和喜达屋集团，建筑师的实践不会被风格或者趋势的改变所影响。他们没有使自己陷入单一的历史风俗或审美风格，而是使他们的设计去追求永恒。

如果詹森·戈德斯坦事务所的建筑物亏欠于现代主义，那么它生来就缺少对所钦佩的20世纪的大师（密斯·凡德罗、埃罗·沙里宁、路易斯·康等）的顺从和敬意，更缺乏对价值观的投入。建立这些现代标志性的建筑，追求开放流动的空间、丰富的自然光、结构的优雅、比例的精准、建筑和内部设计的完美融合。

一般通过材料和纹理的层次感，频繁和为空间制作小部件的手艺人与工匠的深度合作，公司的现代主义风格变得丰富多样。在个人委托的室内装饰项目中，这种特点极其明显，例如240平方米的伦敦阳台阁楼，纽约切尔西附近战前的方块形公寓大楼。其为了重要的时尚管理而建，最终因特色而出现在《建筑设计文摘》的封面上。詹森·戈德斯坦对这个公寓的设计是将黑暗而拥挤的四间卧室改造成明亮而高雅的一间卧室，带有22个新的落地双扇玻璃门，将竖铰链窗对整个弧形阳台开放，使之拥有180度的城市视角。

对内部装饰来说，该公寓唤起了上个世纪20年代的建筑精神。事务所和客户紧密合作，在房主现存的家居和艺术收藏品的周围，设计了一些空间。在起居室，一个带图案的地毯垫在家具下面，家具来自于世界各地的跳蚤市场和古玩店。建筑设计师增加了完全由自己设计的暗色系大理石壁炉架和喷漆的木制品与书架，还给诺尔长凳重新覆盖了个古老的软垫子，为里索姆扶手椅和搁脚凳定做了纺织布，例如灰色的法兰绒和制作晚礼服用的天鹅绒，使之带有纹理。已经完成的空间新旧相结合，既古老又现代，既凉爽又热情。

场所的独特设计解决方案，丰富的保温层材料和与手工艺人的合作，使公司众多的零售工作也颇具特色。曼哈顿肉类加工区的一个精品店，里面是多场地高调时尚的Intermix品牌。事务所为该品牌设计了40多家店，将历史有名的高架索与高架铁路特色融合进部分天花结构之中。在精品店的其他地方，灯光从地面涌入天花板，从街道涌入窗户内，熏过的橡木板从天花板弯曲而下覆盖墙面，在适宜的地方变成地板，创造了吸引客户的舒适木质面。建筑设计师委托纽约当地的贝克·布里顿（Bec Brittain）创造一系列定制的青铜色下垂灯具。灯具的灯光露出绸带状3米高的悬挂幕上的1600块弯曲铝板，使空间在购物者的头上动起来。这种充满回忆的环境，如同伦敦阳台阁楼一样，充满了个人及当地的特色。这些设计特点是实现品牌的目标和客户的梦想，正如住宅为房主所做的一样。

由于事务所向前推进了未来的20年，詹森·戈德斯坦事务所的合伙人继续为老问题——如何使用和占据新空间寻求新的答案。即一个空间传达出的体验和感觉应该是什么样的？一直以来，他们都是在自己严格而深思熟虑的设计流程中寻找解决方案。他们也一直坚信只有拥有对委托的特定环境、情况和愿望的深思熟虑及体贴关怀，才能获得最成功的建筑与室内设计。

商业空间项目
COMMERCIAL WORK

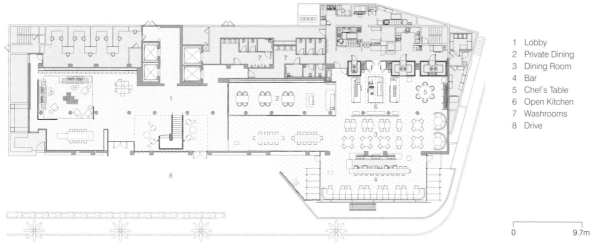

1 Lobby
2 Private Dining
3 Dining Room
4 Bar
5 Chef's Table
6 Open Kitchen
7 Washrooms
8 Drive

Ground Floor Plan

安达仕西好莱坞酒店
Andaz West Hollywood

West Hollywood, CA
2009

Begun in 2004 as a renovation of Los Angeles's Continental Hyatt House (once a Sunset Strip hangout for rock-and-roll royalty), this project became the American debut of Hyatt's Andaz concept, and an entirely new scheme—conceived as an icon for an emerging brand—replaced the original notion. Inspired by the mid-century aesthetic of the 1960s property, the ultimate design pays homage to such masterful L.A. modernists as Richard Neutra and Pierre Koenig. It encloses the balconies of the west-facing rooms in glass to reinvent the building's Sunset Boulevard façade, expanding those rooms into mini-suites with impressive views. A new steel-and-glass pavilion at the base of the hotel's tower, with a site-specific installation by artist Jacob Hashimoto, houses the hotel's restaurant and simultaneously serves as a new marker for the property on Sunset.

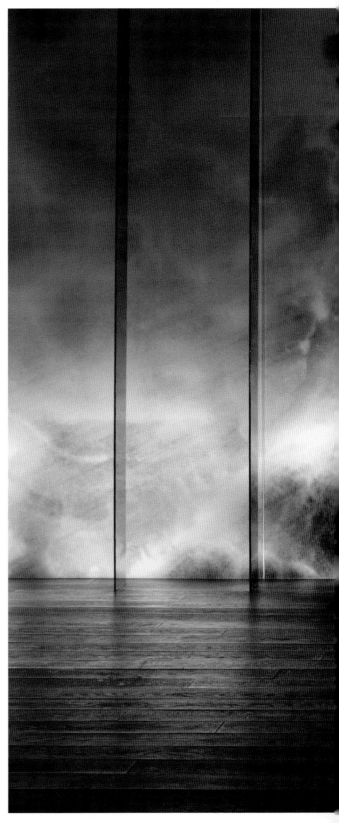

Andaz West Hollywood

Andaz West Hollywood 21

West Hollywood Floor Plan

1 Entry
2 Retail
3 Receiving Area

0   4.8m

阿玛尼家居
Armani Casa

West Hollywood, CA
2001
New York, NY
2001

These stores introduced Giorgio Armani's new Armani Casa home collection to the U.S. market. For the inaugural location, in New York's Soho district, Mr. Armani found inspiration in the neighborhood's distinctive lofts. Their aesthetic and atmosphere translated into a design characterized by open spaces, concrete floors and strong, sculptural shapes—a concept continued at the second shop, in West Hollywood. Both locations use large, prominent skylights to illuminate and give form to spare white-on-white spaces. In New York, a dramatic glass-and-stone staircase sits beneath an angled skylight filling the full width of the store, while in West Hollywood, low walls divide a single level into distinct areas for different product categories, with each zone lit from above by a pyramid-shaped skylight.

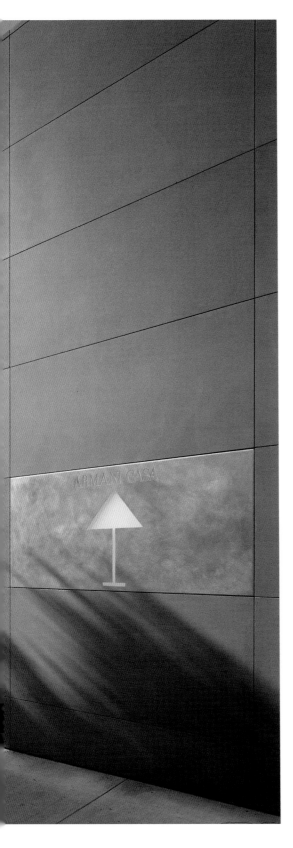

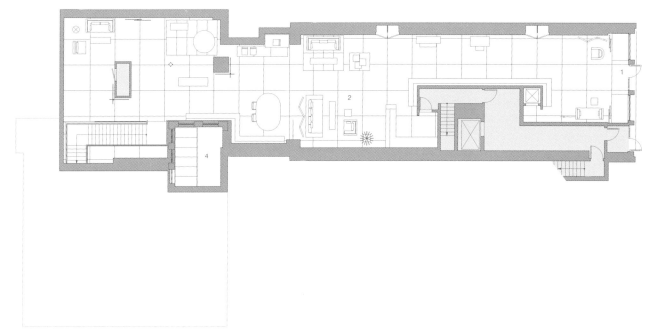

Soho Ground Floor Plan

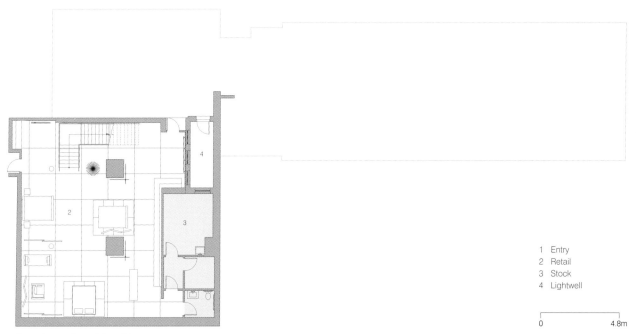

Soho Lower Level Floor Plan

1 Entry
2 Retail
3 Stock
4 Lightwell

0   4.8m

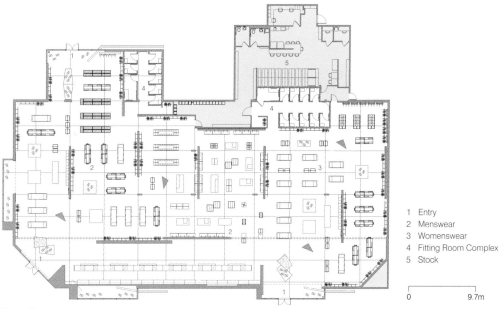

1 Entry
2 Menswear
3 Womenswear
4 Fitting Room Complex
5 Stock

0　　　9.7m

Floor Plan

卡尔文·克莱因公司店
Calvin Klein Company Store

Las Vegas, NV
2013
Harriman, NY
2012

Sunrise, FL
2012
Vaughan, ON
2012

Each approximately 1860 square meters in size and designed as high-volume, high-energy environments, these company stores represent Calvin Klein products in the highly competitive off-price market. The spaces combine men's and women's clothing, previously sold in separate storefronts, as well as footwear and accessories. The new stores make ample use of Calvin Klein's graphic and advertising materials as iconic brand touchstones, their schemes incorporating open ceilings and concrete floors, as well as exposed HVAC and lighting systems. These elements both address the budget considerations inherent in this typology and become emblems of a loft-like, sophisticated atmosphere that appeals to the younger demographic these stores target and serve.

Calvin Klein Company Store

Calvin Klein Company Store

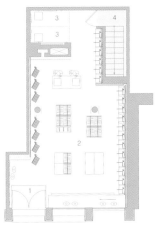

Soho Floor Plan

Hong Kong Floor Plan

1 Entry
2 Retail
3 Fitting Room
4 Stock

卡尔文·克莱因内衣店

Calvin Klein Underwear

Hong Kong
2014
New York, NY
2011
London, UK
2008

This new store model for Calvin Klein Underwear represents a departure from the brand's previously minimalist aesthetic. In this concept, executed worldwide, custom-created sculptural forms house all products, with each form responding to a specific item's packaging and display requirements, as well as its unique features—a sheer intimate, for example, is backlit to highlight its fabric and color. To reduce construction time and cost, the stores consist predominantly of prefabricated elements, with on-site work largely limited to the creation of a neutral background. Otherwise, the sculptural forms containing the merchandise define the design of each space, putting paramount emphasis on the product and resulting in a dynamic environment, even within relatively small confines.

Calvin Klein Underwear

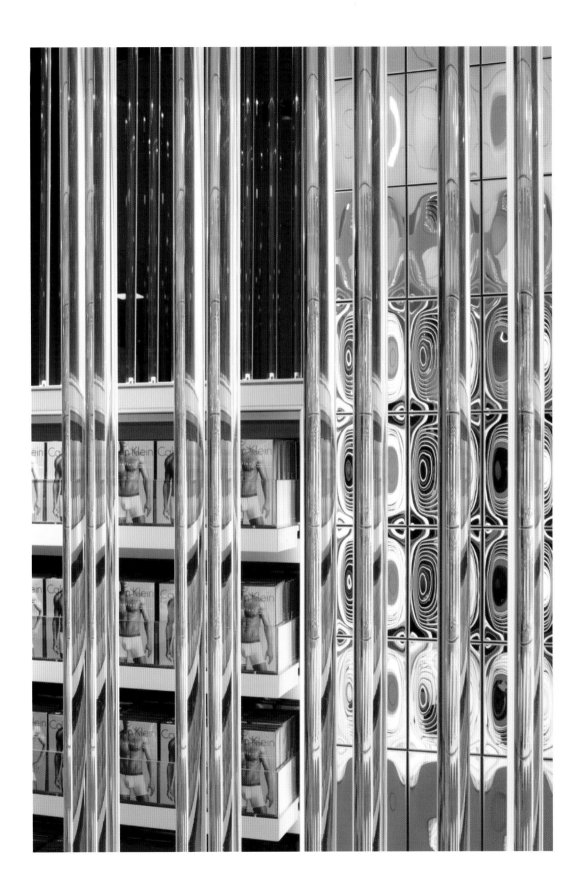

Calvin Klein Underwear

现代海悦酒店
Continental Hyatt House

West Hollywood, CA
2004

Reimagining Los Angeles's Continental Hyatt House—a Sunset Strip home away from home for many oft-misbehaving rock legends—this scheme seeks to recapture some of the property's bad-boy rock-and-roll heritage, letting it live up to its reputation as 'the Riot House.' The design takes as its inspiration a film-noir reinterpretation of the property's 1960s modernist vibe, with a new façade, new public spaces and a new rooftop terrace and swimming pool offering long views of the L.A. basin to the west and the Hollywood Hills to the east. Before beginning this renovation, Hyatt decided to turn the hotel into the first American outpost of its new Andaz brand, and a new iconic concept was developed for that debut. (See page 12.)

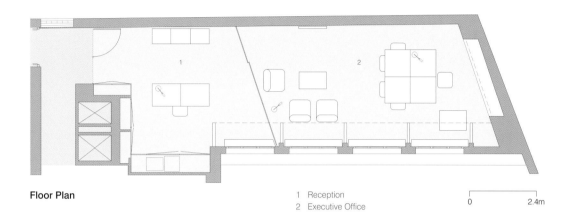

Floor Plan

1 Reception
2 Executive Office

0   2.4m

科蒂集团行政办公室
Coty Inc. Executive Office Suite

Paris, France
2001

This Parisian office suite, created for the chief executive of the French beauty and fragrance company Coty Inc., is a study in minimalism, taking the inspiration for its spare design from the crystal glass of the company's fragrance bottles. The sole components of the office are an executive desk-cum-conference table and a small seating area. Brilliant-white Venetian plaster covers the walls, while white high-tech automotive paint finishes the steel furniture. All the glass used—from the conference table to the walls—is crystal clear and low iron.

Coty Inc. Executive Office Suite 45

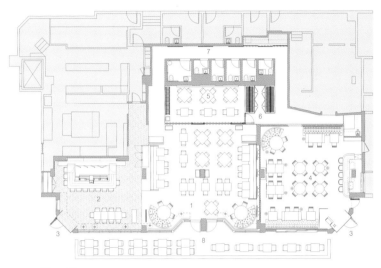

Floor Plan

1 db Bistro Moderne Dining
2 Bar
3 Entry
4 Lumière Dining
5 Private Dining
6 Wine Cellar
7 Washrooms
8 Sidewalk Cafe

0   4.8m

DB现代酒馆
db Bistro Moderne

Vancouver, BC
2008

Created for Michelin-starred chef Daniel Boulud, this project includes the white-tablecloth French restaurant Lumière and more casual db Bistro Moderne. The latter particularly exhibits a deft touch with materials, tones and textures, its inviting atmosphere putting a unique modern spin on the classic Parisian bistro. The bar area features herringbone travertine flooring, a zinc bar top and, behind the bar, woven, polished stainless-steel surfaces framed in red eel skin. Custom handcrafted glass pendants, based on a 1960s Italian design, illuminate the space. Beyond a screen of saw-tooth-textured bronze glass, rolled-steel channeled fixtures illuminate the dining room, where custom-designed chairs of distressed oak and oxblood leather complement banquets in brown- and copper-colored woven leather. A private dining and wine room presents wine racks made of oil-quenched steel and a wall covered in rich red and brown leather tiles.

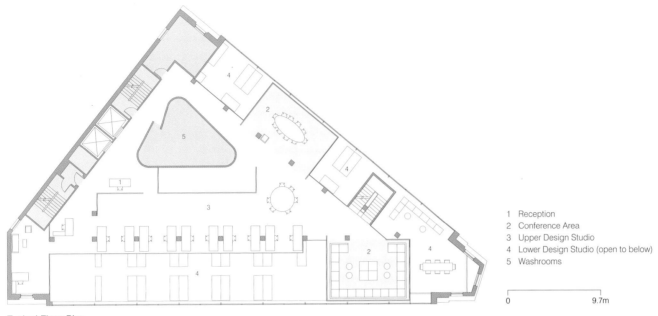

1 Reception
2 Conference Area
3 Upper Design Studio
4 Lower Design Studio (open to below)
5 Washrooms

Typical Floor Plan

犀牛企业总部和展览室
Eckō Unltd. Corporate Headquarters and Showrooms

Perth Amboy, NJ
2004

Eckō Unltd. commissioned the conversion of a 12-story, Depression-era landmark in Perth Amboy, NJ, into new headquarters. A retail space occupies the base, with offices, showrooms and design studios above, plus a new penthouse with gym, locker rooms and a half-size basketball court. Tasked by the urban fashion brand to develop shared, hierarchy-free workspaces, but required by the existing structure to divide these spaces across three floors, the new scheme suggests a unique solution: it removes perimeter floor plates and clads upper levels in glass to create a shared, three-level environment around the building's edges. Cantilevered glass boxes within this atrium house showrooms and meeting spaces, while a glass-enclosed stair tower connects to the penthouse, itself clad in polished stainless steel to reflect, and blur with, the sky.

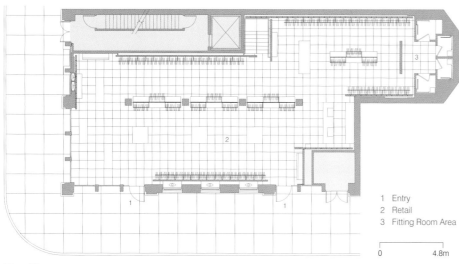

Floor Plan

1 Entry
2 Retail
3 Fitting Room Area

0   4.8m

爱姆普里奥·阿玛尼苏活区店
Emporio Armani Soho

New York, NY
2002

In contrast to the series of Emporio Armani stores throughout the U.S. designed in a single common idiom, this location stands as a one-of-a-kind environment. Although the space is modest in size, the brand saw it as a flagship, stocking a spare, curated assortment. Mr. Armani's concept for the store, which sits on a prominent corner in New York's Soho, sought to represent the unique vibe of its downtown location and customer base. To that end, high and low, old and new, rough and refined sit in poignant juxtaposition here: a honed-granite floor abuts sandblasted concrete-block walls, for example, against which sit the store's sophisticated offerings. At the center of the space, frosted acrylic, lit from within, surrounds original cast-iron columns—a ghost of the past seen through a modern shroud.

Emporio Armani Soho

Typical Duplex Floor Plan

1 Elevator Lobby
2 Living Room
3 Kitchen / Dining Area
4 Media / Library
5 Powder Room
6 Bathroom

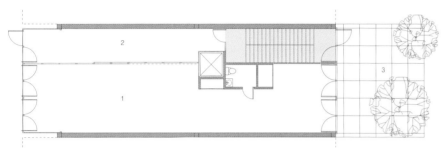

Ground Floor Plan

1 Retail Tenent
2 Condominium Entrance
3 Garden

熨斗区豪华阁楼

Flatiron Luxury Loft Residences

New York, NY
1999

This mixed-use, luxury residential and retail project, which rises mid-block between Sixth and Seventh Avenues in New York's Flatiron District, evokes the warmth and clean lines of Isamu Noguchi's 1950s Akari light sculptures. It juxtaposes a two-story retail base and duplex condominium apartments above, both sheathed in bronze-toned glass, with one-level condo units that each have a terrace and a recessed clear-glass wall. By alternating the double-height apartments with those of a single story, a series of lantern-like illuminated stacked cubes emerges. Inside, the loft-like apartments' clean, open plans allow for easy flow from front to back, with a large, double-height space in each duplex facing the street. Residents share a ground-level garden and a rooftop terrace.

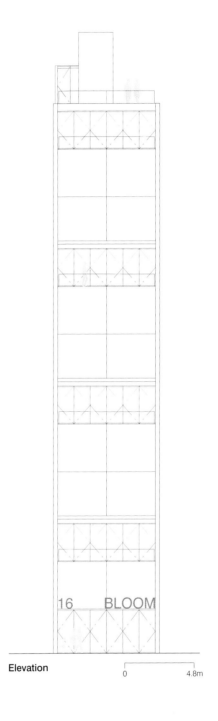

16 BLOOM

Elevation 0 4.8m

Flatiron Luxury Loft Residences 59

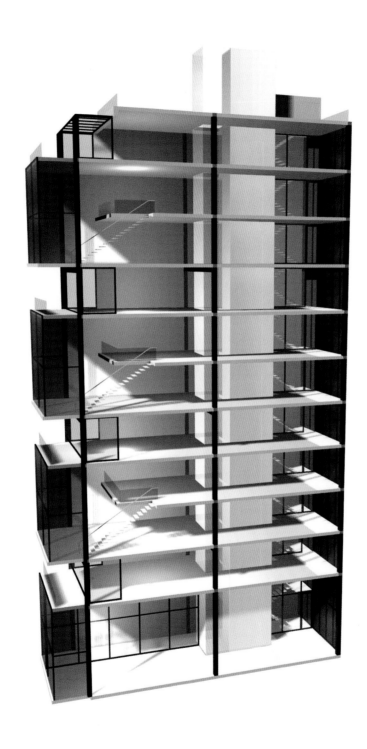

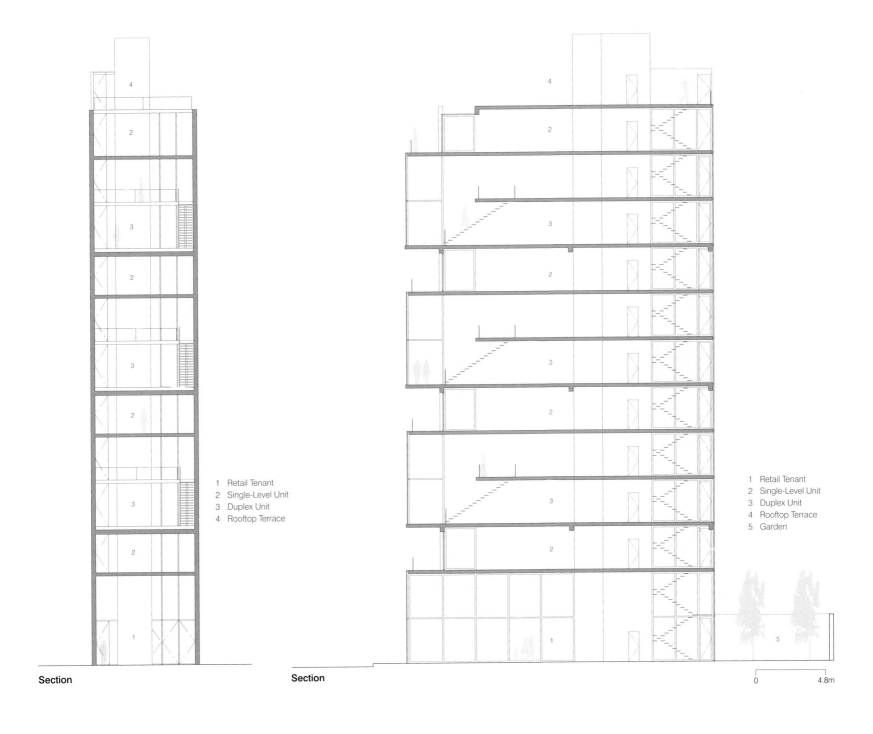

Flatiron Luxury Loft Residences

Vancouver Site Plan
1 Holt Renfrew
2 Pacific Center
0 — 61m

霍尔特伦弗鲁旗舰店
Holt Renfrew Flagships

Toronto, ON
2013
Calgary, AB
2009
Vancouver, BC
2007

The continuation of a long-term partnership dating to 2005, these new concepts for Canadian luxury retailer Holt Renfrew's three flagships share design DNA but are adapted to their unique settings. The first, in Vancouver, breaks long-established department-store tropes, abandoning the lifeless, windowless brick box with interiors defined by 'hard' aisles (stone) and 'soft' pads (carpet). Instead, high-tech glass clads the exterior, and luminous white-marble floors unify light-filled interiors. Created with Vancouver's Nathan Allan Glass Studios, the iconic 'pillowed' glass panels have a quilted appearance, their convex cells refracting light differently throughout the day. At the building's heart, a three-story atrium sits beneath a giant skylight as a dynamic public attraction, connecting the 12,000-square-meter space's shopping levels. The Calgary flagship adapts this successful formula, anchoring downtown's Core development.

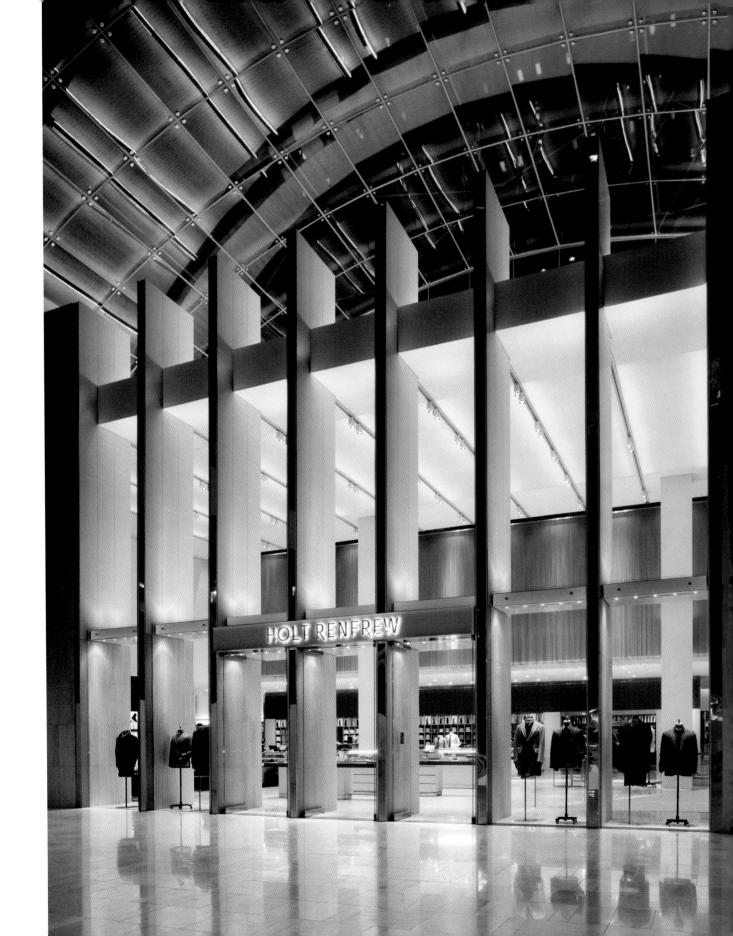

Holt Renfrew Flagships

霍尔特伦弗鲁旗舰店——新一代
Holt Renfrew Flagship – Next Generation

Mississauga, ON
2016

This freestanding 11,150-square-meter store in the West Toronto suburb of Mississauga represents the next generation in large-scale luxury retailing for the high-end Canadian department store Holt Renfrew. Sheathed entirely in an iconic and proprietary glass façade created exclusively for Holt Renfrew by Janson Goldstein, the structure represents a key element of downtown Mississauga's new master plan. The building stands as a glowing glass box whose scale and prominence equal the area's other major landmarks, including a performing arts center and city hall, which sit directly opposite. The flagship sets a new standard for department-store design, employing natural daylight to illuminate the interiors of its entire perimeter, and allowing visual communication between indoors and out.

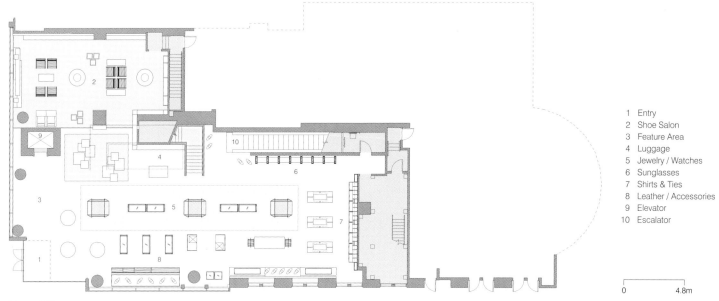

Ground Floor Plan

1	Entry
2	Shoe Salon
3	Feature Area
4	Luggage
5	Jewelry / Watches
6	Sunglasses
7	Shirts & Ties
8	Leather / Accessories
9	Elevator
10	Escalator

霍尔特伦弗鲁男士精品店

Holt Renfrew Men

Toronto, ON
2014

Holt Renfrew's debut stand-alone men's concept store, this 1500-square-meter space on Toronto's fashionable Bloor Street represents a departure for the Canadian luxury retailer, its masculine design resembling a private club as much as a store. On the main level, internally lit shelving in gray-stained white oak and blackened stainless steel surrounds central columns. A custom mosaic-tile floor connects leather goods, jewelry and men's furnishings, while a hand-knotted custom carpet and deep-blue, leather-upholstered, polished-stainless-steel chairs define the footwear area. On the fashion-focused second level—which includes a personal-shopping suite and bespoke-tailoring shop—walnut shutters and herringbone flooring create a residential feel, and a patterned-wool runner marks a circulation path. A double-height structural screen extends through the escalator atrium, creating a strong visual connection between the two floors and providing additional visual-merchandising space.

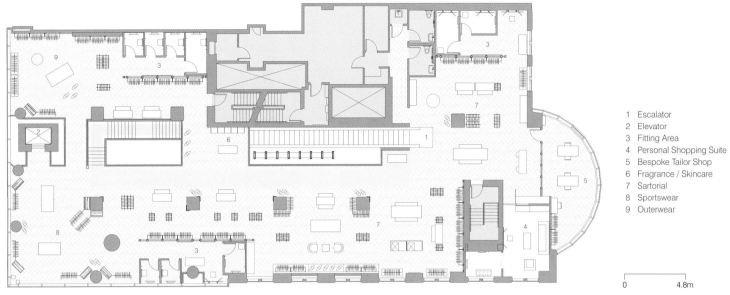

Second Floor Plan

1 Escalator
2 Elevator
3 Fitting Area
4 Personal Shopping Suite
5 Bespoke Tailor Shop
6 Fragrance / Skincare
7 Sartorial
8 Sportswear
9 Outerwear

0   4.8m

Holt Renfrew Men 85

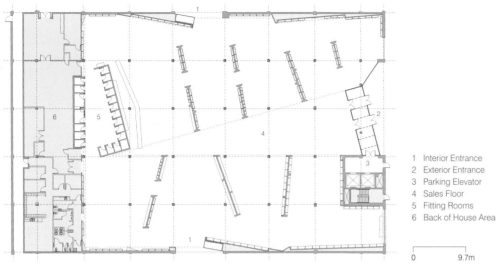

Floor Plan

1 Interior Entrance
2 Exterior Entrance
3 Parking Elevator
4 Sales Floor
5 Fitting Rooms
6 Back of House Area

卡尔捷迪克斯区hr2品牌店
hr2 Quartier DIX30

Montreal, QC
2013

Designed to introduce Canadian luxury retailer Holt Renfrew's off-price concept, this store stands as a bold brand statement. It occupies the terminus of the central axis of Montreal's Quartier DIX30 development, the facets of its perforated aluminum façade creating a striking sculpture that both captures and reflects daylight and evening illumination. At night, it also glows warmly from within, becoming a canvas for a computerized LED display. The interior proceeds from the forms of the exterior, split into men's and women's halves by a central spine connecting the façade to the cash/wrap opposite. In between, customers browse a series of 'hedges'—3-meter-high walls that both house merchandise and separate different product categories, just as boxwoods divide a garden—then return down the spine to exit through the iconic, light-filtering façade.

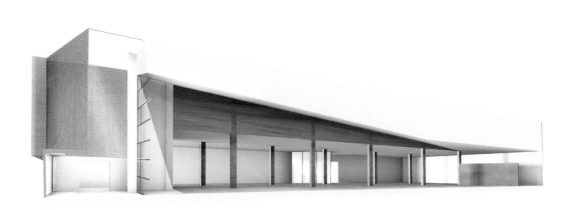

hr2 Quartier DIX30

Intermix品牌店
Intermix

Madison Avenue, New York, NY
2014
Soho, New York, NY
2013

Brooklyn, New York, NY
2013
Bowery, New York, NY
2013

Since 2010, Janson Goldstein's collaboration with Intermix has resulted in 40-plus boutiques for the multi-vendor, high-end fashion retailer. A brand-wide visual language unites them, but each store's neighborhood context provides distinct inspiration. For example, on New York's Bowery, a former deli combines elements of the area's bygone punk-rock days (wire-brushed brick, iron-beam racks) with sophisticated furnishings referencing its current fashionable status. In a once industrial and now family-friendly stretch of Brooklyn's Carroll Gardens, a rebuilt brick storefront and largely original interiors embrace the street with a new oversized display window and entry. On Madison Avenue, screens made of darkened bronze, blackened steel, woven bronze and stainless steel reference the Upper East Side's reputation for luxury, while Soho evokes an iconic downtown artist's loft, incorporating cast-iron columns and sculptural elements.

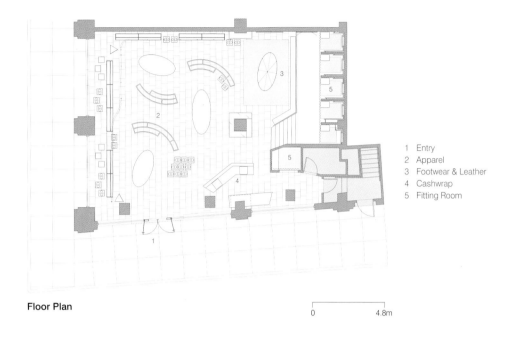

1 Entry
2 Apparel
3 Footwear & Leather
4 Cashwrap
5 Fitting Room

Floor Plan

0　　4.8m

Intermix肉类加工区品牌店
Intermix Meatpacking

New York, NY
2011

The first of a new generation of stores designed for the multi-vendor, high-fashion retailer Intermix, this 230-square-meter boutique in Manhattan's Meatpacking district nestles beneath the High Line, the park on the West Side's 1930s-era elevated railroad. In the soaring space, a 3-meter-tall, ribbon-like screen made of 1600 panes of curved, mirror-polished aluminum hangs from the steel track, diffusing and reflecting daylight from floor-to-ceiling windows and site-specific light installations by Brooklyn designer Bec Brittain. In the fitting area, fumed European white oak wraps the floor, walls and ceiling, creating a wooden grotto. Elsewhere, an oversized custom-designed oval ottoman covered in soft, pale-purple leather serves as a central seating element in the footwear area, and antiqued, slightly rusticated black Belgian limestone covers the floor in a random pattern, referencing the neighborhood's cobblestone streets.

Intermix Meatpacking

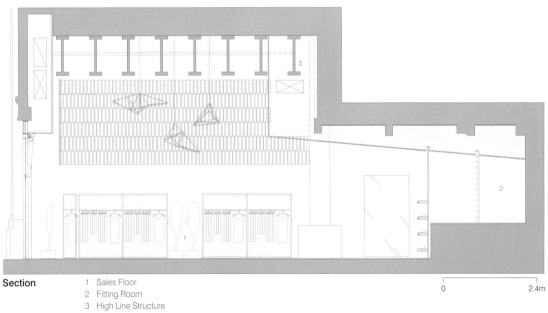

Section
1 Sales Floor
2 Fitting Room
3 High Line Structure

0　　2.4m

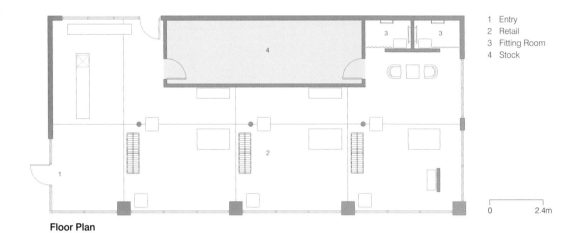

Floor Plan

1 Entry
2 Retail
3 Fitting Room
4 Stock

KBond男装精品店
KBOND

Los Angeles, CA
1999

Owned by, and designed in collaboration with, installation artist Karen Kimmel and her husband, James Bond, this innovative Los Angeles menswear boutique occupied a glass-fronted raw space clearly visible to pedestrians and cars on Beverly Boulevard, especially at night, when it was illuminated from within. Mindful of the brand's desire to integrate artistic interventions, and working with a limited budget, the scheme for the store took creative advantage of existing conditions. Bold but simple sculptural constructions made of theater scrim and gelled fluorescent lighting both divided the space and displayed hanging products, while glass-topped architectural flat files housed folded pieces. Opposite the glazed façade, a 9-meter-long wall became not only a backdrop for the merchandise, but also a blank canvas that a different artist-collaborator would refresh with a new installation every 60 days.

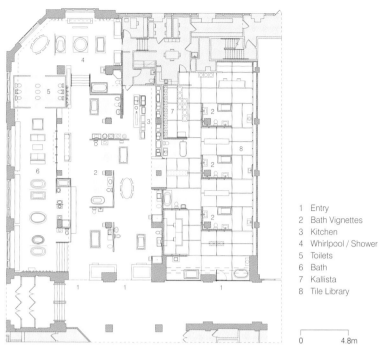

1 Entry
2 Bath Vignettes
3 Kitchen
4 Whirlpool / Shower
5 Toilets
6 Bath
7 Kallista
8 Tile Library

0   4.8m

Floor Plan

科特 / 安萨克斯店
Kohler / Ann Sacks

Chicago, IL
2006

Occupying adjacent spaces in Chicago's famed Merchandise Mart, these interrelated projects represent Kohler's first-ever direct-to-consumer retail space, and a new showroom type for Ann Sacks. To maximize its ability to communicate directly with customers, the Kohler space takes the form of a design exhibition devoted to the company's kitchen and bath fixtures, with a clear circular path leading through the individualized galleries created for each product line. As such, it serves multiple functions: brand bastion, retail outlet, product showroom. The design for Ann Sacks, meanwhile, which is also a Kohler company, presents the tile maker's full range of products. Here, the space functions as a sort of library, with a blackened-steel shelving system, and mosaic flooring that highlights the tiles as they might be used in customers' own homes.

Kohler / Ann Sacks

Kohler / Ann Sacks 111

路易斯登集团办公室
LVMH Corporate Offices

New York, NY
2011

French fashion and luxury-goods group LVMH commissioned this trio of wall reliefs to mark the entry points of its corporate offices in New York City. All three pieces share the same overall design—a gridded, multilevel installation based on the circular quatrefoil motif of the signature Louis Vuitton pattern, rendered at various scales. Each was crafted in a different material, with the wood, plaster, and polished-metal versions used to differentiate one floor from another in the office's elevator lobbies. This, in turn, eases wayfinding while also providing a strong brand identity.

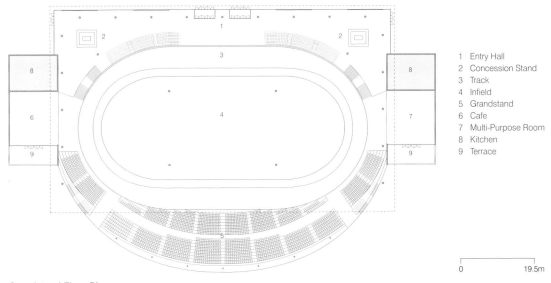

Grandstand Floor Plan

1 Entry Hall
2 Concession Stand
3 Track
4 Infield
5 Grandstand
6 Cafe
7 Multi-Purpose Room
8 Kitchen
9 Terrace

纽约奥林匹克自行车馆
New York Velodrome

New York, NY
2009

VoNYC Inc.—a privately funded non-profit promoting track cycling in New York City—commissioned a design concept and feasibility study for this 3000-seat, 10,200-square-meter multi-use building, which combines athletic facilities with retail space, a cafe, a community center, exhibition galleries, public art, offices, locker rooms, and pre-function areas. Designed for a variety of cycling competitions, the Velodrome's 200-meter track surrounds a 2040-square-meter infield that can be used for other sports, concerts, film screenings, and conferences. A continuous glass curtain wall surrounds the track and seating areas, defining the space and allowing the public to look in, and sports spectators to gaze out. The wooden track and its structure sit like sculptures within this glass box, while a green roof with solar panels tops the building, designed to achieve LEED Gold certification.

蒙特利尔奥美集团
Ogilvy Montreal

Montreal, QC
2014

This feasibility study considers the renovation and rebranding of a historic 25,550-square-meter department store in Montreal's heart, including a 4650-square-meter addition adjacent to the original 1896 building. The proposal weaves the 21st century into the 19th, connecting all six floors across both buildings, modernizing and activating the original while embracing its heritage and sense of place. Using architectural innovation to create inviting destinations, the scheme centers on a new atrium containing primary vertical circulation as well as the 'Digital Concierge'—a six-story LED-wrapped column that serves as both art and wayfinding signage. A massive contemporary skylight caps the atrium, infusing the once-dark interiors with daylight and attracting customers to higher floors. This glass structure projects across the rooftop and beyond the historic building's façade, creating an illuminated beacon, an icon in both the streetscape and the skyline.

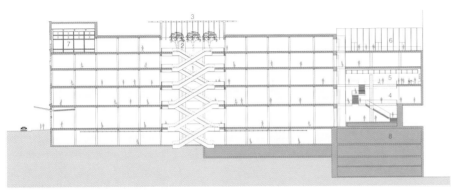

1	New Atrium
2	Digital Concierge
3	New Skylight
4	Men's Townhouse
5	Brasserie
6	Terrace
7	Tudor Room
8	Parking

0　　9.7m

Section

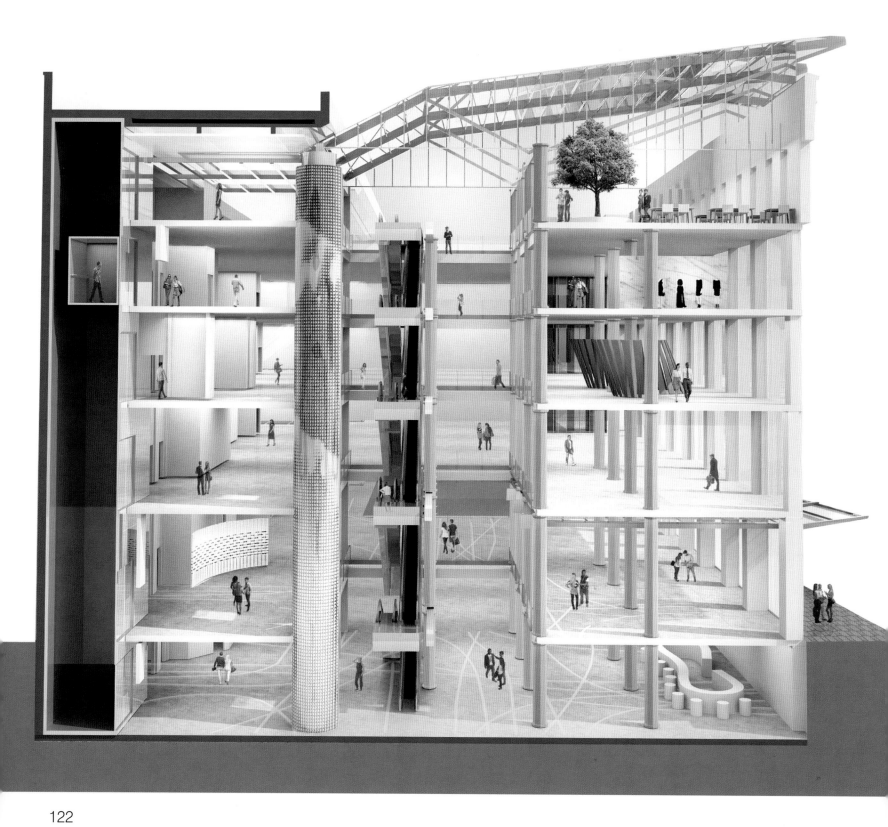

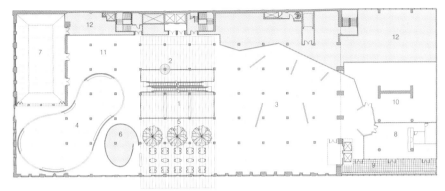

Fifth Floor Plan

1 New Atrium
2 Digital Concierge
3 Designer Footwear Salon
4 Contemporary Footwear Salon
5 Cafe / Event Space
6 Kitchen
7 Tudor Room
8 The Apartment
9 Terrace
10 Personal Shopping
11 Market
12 Back of House

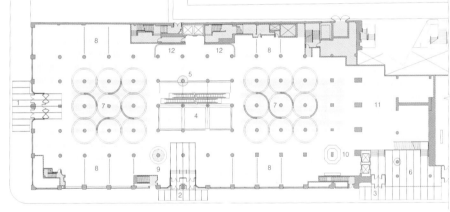

Ground Floor Plan

1 Sainte-Catherine Street Entrance
2 de la Montagne Street Entrance
3 Men's Townhouse Entrance
4 New Atrium
5 Digital Concierge
6 Men's Townhouse
7 Handbags / Leather Hall
8 Vendor Shops
9 Florist
10 Patisserie
11 Jewelry & Watches
12 Eyewear

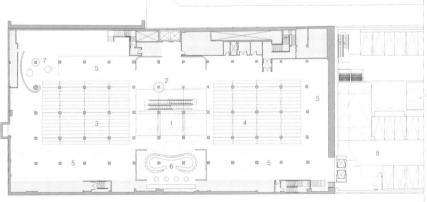

Concourse Floor Plan

1 New Atrium
2 Digital Concierge
3 Color Pavilion
4 Skincare Pavilion
5 Vender Shops
6 Juice Bar / Grab & Go
7 Beauty Services
8 Valet Parking

Ogilvy Montreal

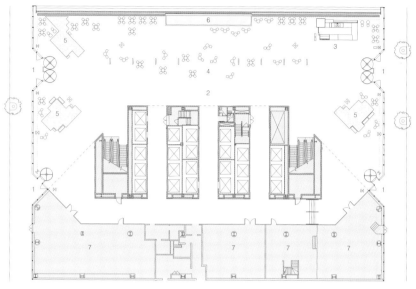

Floor Plan

1 Entry
2 Public Arcade
3 Cafe
4 Seating Area
5 Bamboo Planters
6 Fountain
7 Retail

公园大道广场公共中庭
Park Avenue Plaza Public Atrium

New York, NY
2015

This redesign of a prominent Midtown Manhattan office tower's 2600-square-meter public atrium and lobby lets the building live up to the architectural pedigree of its neighborhood, which includes such modernist icons as the Seagram Building. The new scheme centers on a dramatic glass colonnade composed of eight sculptural, round-edged piers, each nine meters high, one meter wide, and 15 centimeters thick. This colonnade pulls the public in, making a path through the space, providing places to gather and offering rhythm and order. The columns also encourage spontaneity and a variety of interactions, with movable seating and a free-flowing plan allowing for multiple uses. Art, sculpture, music, digital media, and a central water feature create diverse experiences, while a continuous green wall and bamboo planters suggest an urban Eden.

Park Avenue Plaza Public Atrium

Park Avenue Plaza Public Atrium

Rocket Dog 集团总部
Rocket Dog Brands Corporate Headquarters

Los Angeles, CA
2012

Occupying the second and third floors of a corner building in Los Angeles's fashion district, this 1020-square-meter space provides executive offices, design studios, and a West Coast showroom for the multi-brand footwear company. Core to the scheme is the third-floor courtyard—accessible from all sides and flanked by the design studio and a cafe—which offers natural light and fresh air as well as a central gathering place. The third floor also houses the CEO's office, while the second contains additional workspace and the showroom. A dramatic steel-and-marble staircase connects the two floors, leading from reception to the courtyard. Selected to evoke the beach, in particular nearby Malibu, the palette features distressed oak and glass in the offices, with teak-and-steel lattice enhancing the courtyard's plaster walls.

Upper Level Floor Plan
1. Design Studio
2. Office
3. Conference Room
4. Courtyard
5. Cafe
6. Library
7. Washroom

Lower Level Floor Plan
1. Reception Area
2. Showroom
3. Conference Room
4. Office / Open Plan
5. Washroom

0   4.8m

Rocket Dog Brands Corporate Headquarters

萨克斯第五大道精品百货店
Saks Fifth Avenue

Sarasota, FL
2014
San Juan, PR
2014
Naples, FL
2009

Boca Raton, FL
2006
Beverly Hills, CA
2006

Part of a long relationship with Saks Fifth Avenue, this series of stores represents a new generation of shopping environments, one designed to emphasize the brand's 90-year heritage as an iconic luxury brand while simultaneously demonstrating its focus on the future. Ground-up projects in San Juan, Puerto Rico, and Sarasota, FL, as well as renovations of locations in Beverly Hills and South Florida, prove particularly emblematic. The San Juan and Sarasota stores feature bold façades inspired by mid-20th-century tropical modernism. Plant-covered living stonewalls at ground level give way to glazing shaded by bronze-toned *brise-soleil* above, creating a striking and iconic brand profile. Inside each location, natural sunlight gently filters through the *brise-soleil* into the store's most special spaces, including a sumptuous cafe and private-shopping suite.

Saks Fifth Avenue

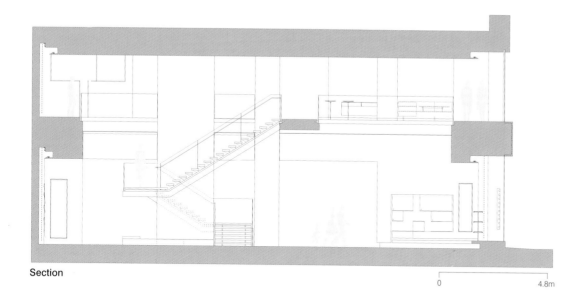
Section 0 4.8m

菲拉格慕旗舰店
Salvatore Ferragamo

Fifth Avenue Flagship, New York, NY
2003
Manhasset, NY
2002

The flagship store is the central element of a multi-year collaboration with the Italian fashion house that resulted in new locations across the US. The design of this 1900-square-meter, bi-level flagship on Manhattan's Fifth Avenue inaugurated a new generation of stateside Salvatore Ferragamo stores. The flagship seamlessly melds two existing retail spaces across a pair of prominent buildings, creating a singularly grand Ferragamo environment. Containing men's and women's collections, state-of-the-art fitting rooms, a custom-footwear atelier and a gallery, the space draws customers in with enticing views down its bays, and such dramatic features as a wood, stone, and glass staircase that wraps a monolithic stone column. This terminates in a bridge connecting the store's east and west sides, leading to the men's footwear salon in a glass-enclosed aerie overlooking Fifth Avenue.

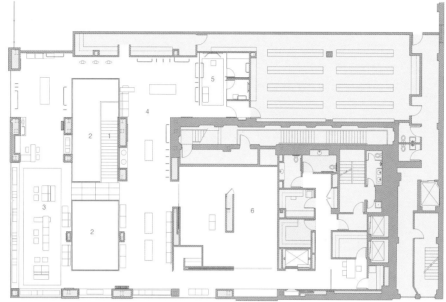

1 Stair
2 Open to Below
3 Men's Footwear
4 Men's Apparel
5 Fitting Area
6 Gallery

Second Floor Plan

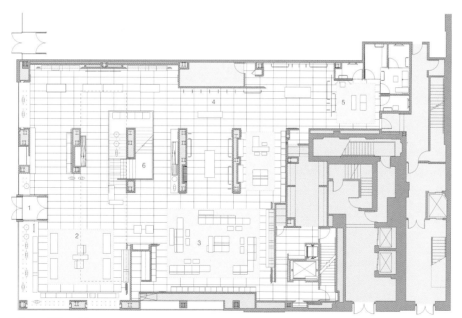

1 Entry
2 Handbags
3 Women's Footwear
4 Women's Apparel
5 Fitting Area
6 Stair

Ground Floor Plan

0   4.8m

Salvatore Ferragamo

Floor Plan

1 Hotel Entrance
2 Mix
3 Courtyard
4 Cafe
5 Retail

0   9.7m

浪花棕榈滩零售店
The Breakers Palm Beach Retail Shops

Palm Beach, FL
2008

Reimagining the glass-enclosed central loggia of The Breakers Palm Beach, this new master plan for the property's retail courtyard elevates and expands the shopping experience at the historic Italian Renaissance–style hotel. The concept includes a fully realized fashion-jewelry boutique called The Mix, and the development of additional branded boutiques for clothing, swimwear, and footwear. These elements enhance the guest experience, generate significant revenue, and establish The Breakers as a retail destination for affluent locals, benefiting both retail and dining venues. Freestanding at the courtyard's focal point, The Mix serves as a beacon. Its four sides of 4-meter-high, classically inspired glass arches, and unique, vertical, eye-level window displays show the jewelry off to maximum effect, especially at night, when the store's glowing windows become a major draw.

The Breakers Palm Beach Retail Shops   147

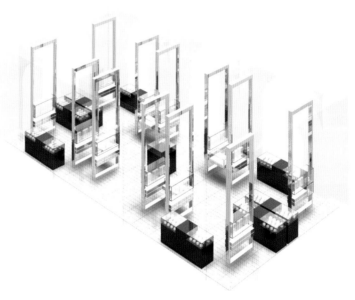

The Breakers Palm Beach Retail Shops

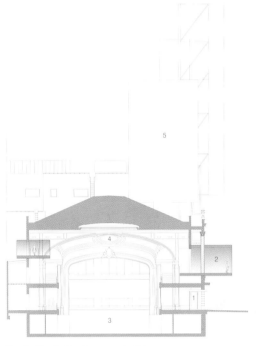
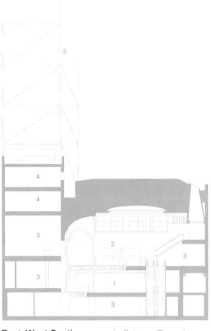

North–South Section
1 Entry
2 Glass Marquee
3 Retail
4 Historic Proscenium
5 Schematic Signage Superstructure

East–West Section
1 Entrance Beyond
2 Glass Marquee Beyond
3 Retail
4 Office (Former Flyloft)
5 Schematic Signage Superstructure

时代广场改造性再利用
Times Square Adaptive Reuse

New York, NY
2005

The streetwear label Eckō Unltd. commissioned an extensive feasibility study to redevelop Manhattan's historic Times Square Theater, the only Broadway playhouse not yet repurposed by The New 42nd Street, an organization overseen by New York's Empire State Development authority. The study examines the adaptive reuse of the theater, considering its conversion into a flagship retail space and what Eckō calls an 'urban youth epicenter.' The proposal's most compelling feature is its reinterpretation of the marquees that once lined the street; to preserve the landmarked limestone façade and simultaneously advertise the building's new use, an occupiable glass box was designed to project through the building's colonnade and over the sidewalk. Inside the building, the auditorium becomes a grand, multi-level retail space delicately woven into the preserved historic structure and soaring proscenium.

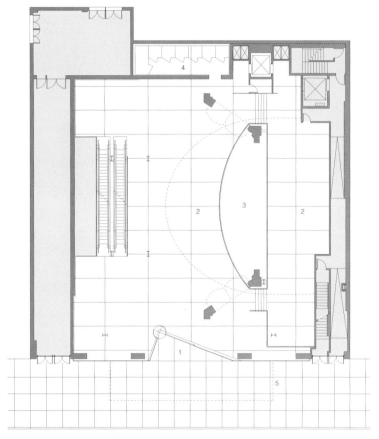
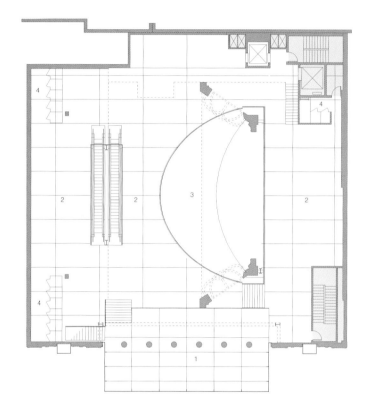

Ground Floor Plan
1. Entry
2. Retail
3. Open to Below
4. Fitting Room Complex
5. Glass Marquee Above

Second Floor Plan
1. Glass Marquee
2. Retail
3. Open to Below
4. Fitting Room Complex

0 — 9.8m

Times Square Adaptive Reuse

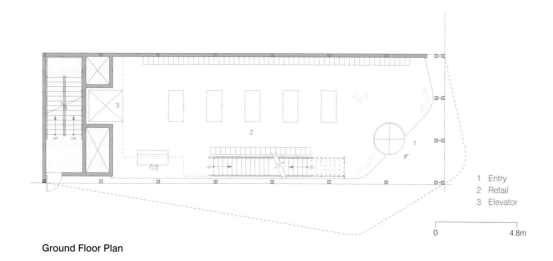

Ground Floor Plan

时代广场商业开发
Times Square Commercial Development

New York, NY
2014

The feasibility study for this site, located at the north end of Manhattan's Times Square, envisioned a new, single-tenant, 740-square-meter flagship location for a large retail brand. Offering four stories of retail space and one floor of offices and showrooms on what is currently an underutilized, single-level site, the new structure maximizes the allowable floor-area ratio (FAR), as well as retail square footage, visible street frontage, and space for LED billboard advertising on the building's façade—an iconic element and a prime revenue driver of any new Times Square development.

Times Square Commercial Development

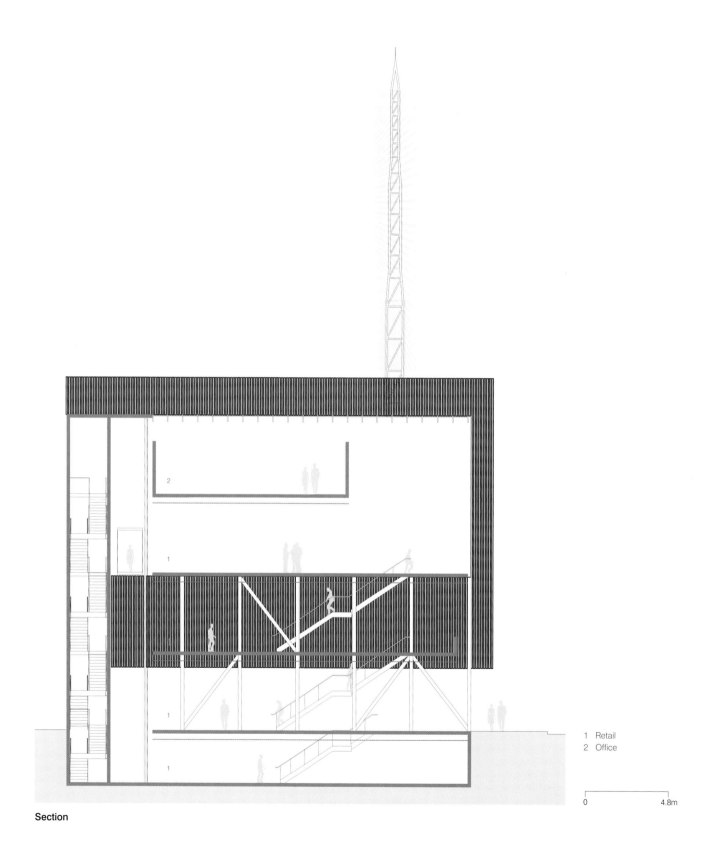

Section

1 Retail
2 Office

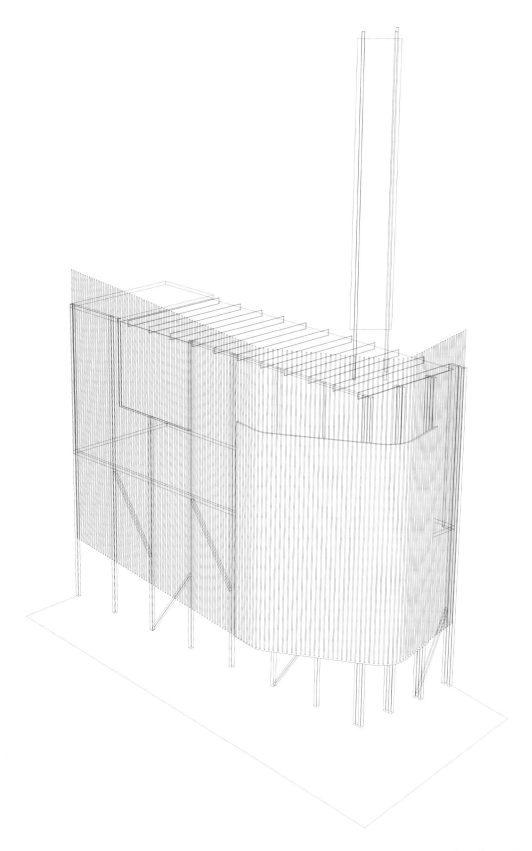

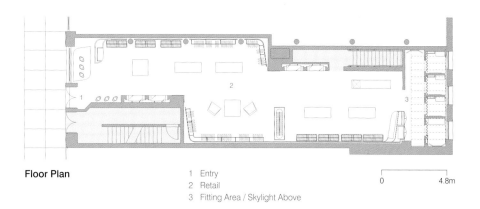

Floor Plan
1 Entry
2 Retail
3 Fitting Area / Skylight Above

0   4.8m

苏活区TSE品牌店
TSE Soho

New York, NY
2008

The design for this downtown Manhattan outlet of the high-fashion cashmere label TSE was born out of the aesthetic of a historic, quintessentially gritty Soho loft, now newly outfitted with modern, luxurious finishes. The boutique retains the brick walls, oak floors, and exposed ductwork and piping of the original raw space, juxtaposing them with refined, highly lacquered, soft-white wall panels whose curving shape creates a ribbon effect throughout the store. The smooth finishes of these panels both contrast with and complement the rich, intricate textures of the cashmere. Lightweight display fixtures—made of gently arched, satin-finished stainless-steel bars—hang from the ceiling or stand in front of walls, giving the sense that the clothing is floating above the floor or in front of the lacquered panels, rather than simply hanging.

Bloom Floor Plan

1 Entry
2 Flower Display
3 Reflection Pool
4 Home Furnishings
5 Arrangement Area
6 Cooler

莱克星顿大街W酒店
W Hotel Lexington Avenue

Bliss 49 + Spa Suites, New York, NY
2005
Bloom Flowers, New York, NY
2000

This floor-through renovation of Starwood Hotels & Resorts' original W Hotel encompasses an expansive 2300 square meters, including a spa, a gym, men's and women's lounge and locker facilities, three retail spaces, 11 guest rooms, and four spa suites specifically designed for guests focusing their stays on health and wellness. On the northeast corner of the street level, a satellite of the well-known New York florist Bloom puts the theatricality of flower arranging at center stage, giving customers a window into the artistry. Adjacent to Bloom is the entry for the Bliss 49 Spa, as well as its salon and retail outlet. Just beyond lie dedicated elevators that deliver clients directly to the fourth floor, which is exclusively devoted to spa services and spa suites.

W Hotel Lexington Avenue

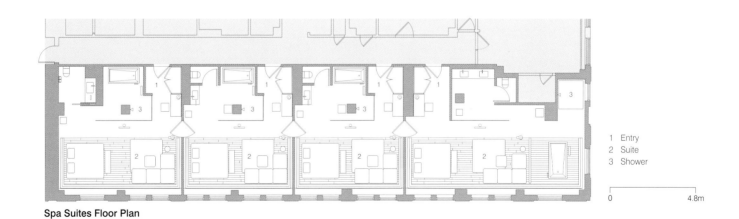

Spa Suites Floor Plan

1 Entry
2 Suite
3 Shower

0　　4.8m

五星级酒店与住宅
5 Star Hotel and Residences

New York, NY
2006

This mixed-use project near Lower Manhattan's historic South Street Seaport comprises a 50-story condominium tower and a six-floor base structure housing a five-star hotel. Creating an intimate, luxurious arrival experience, a unique porte-cochère allows guests and residents to drive through the building, removing them from street and pedestrian traffic. From the porte-cochère, the condominium lobby sits to the east, and the hotel's lobby, restaurant and bar to the west, all with 6-meter-high ceilings, and façades sheathed in glass. Atop the hotel, a landscaped terrace includes a gym, a yoga studio, and a lap pool perched on the building's edge, overlooking the Brooklyn Bridge, while the tower's rooftop features a swimming pool enclosed by a 6-meter-high glass wind screen. The condominiums' loft-like floor plans emphasize expansive river and harbor views.

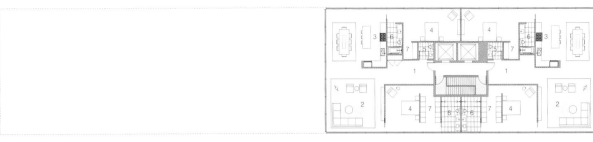

Two Unit Floor Plan

1 Gallery
2 Living Room
3 Kitchen / Dining
4 Bedroom
5 Powder Room
6 Bathroom
7 Walk-in Closet

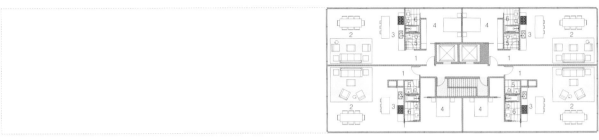

Four Unit Floor Plan

1 Entry
2 Living / Dining
3 Kitchen
4 Bedroom
5 Powder Room
6 Bathroom

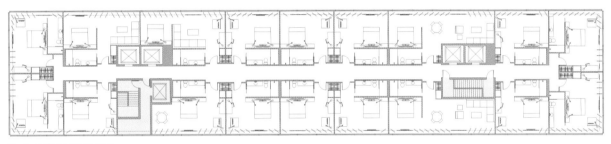

Typical Hotel Floor Plan

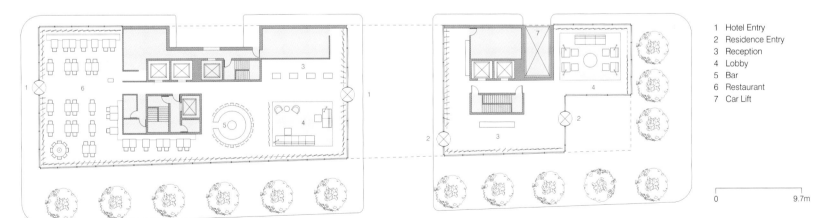

Ground Floor Plan

1 Hotel Entry
2 Residence Entry
3 Reception
4 Lobby
5 Bar
6 Restaurant
7 Car Lift

0   9.7m

5 Star Hotel and Residences   171

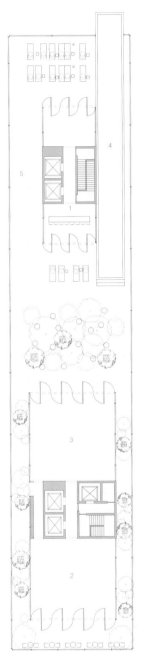

1 Juice Bar
2 Yoga Studio
3 Gym
4 Lap Pool
5 Terrace / Garden

0   9.8m

Roof Plan

5 Star Hotel and Residences

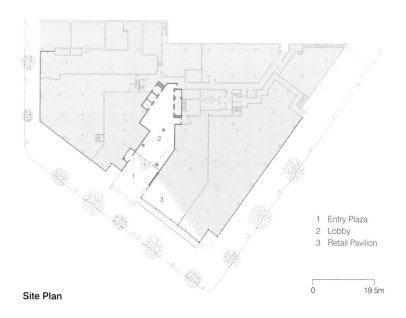

1 Entry Plaza
2 Lobby
3 Retail Pavilion

Site Plan

华盛顿新罕布什尔大道1200号
1200 New Hampshire Avenue NW

Washington, DC
2012

Near the Washington, DC, neighborhood of Dupont Circle, the addition of a new entry plaza, lobby entry and, most prominently, a sparkling glass retail pavilion transforms a 35-year-old brick office building. Carefully inserted into the foreground of the existing structure, the 929-square-meter pavilion is clad in custom-made, 6.7-meter-high glass panels. These panels feature a patterned, mirrored interlayer that both reflects the streetscape—including a small park across the street—and allows for views into the new space, now occupied by a bank and a future television studio. The design of the glass and its uninterrupted, mullion-free installation abstract the pavilion's mass, and shifting natural light creates continuous changes in its appearance. Inside, the lobby combines a crystal-clear, low-iron glass façade, oil-quenched-bronze walls, flame-finished granite and polished quartzite.

1200 New Hampshire Avenue NW

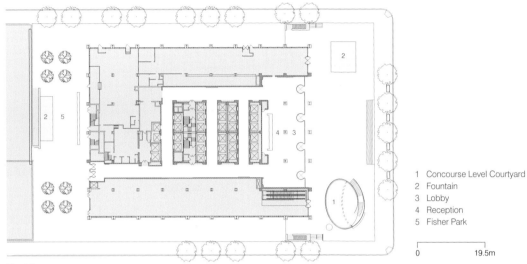

1 Concourse Level Courtyard
2 Fountain
3 Lobby
4 Reception
5 Fisher Park

0   19.5m

Site Plan

美洲大道1345号
1345 Avenue of the Americas

New York, NY
2014

This project calls for the reinvigoration of the 50-story Midtown Manhattan office tower at 1345 Avenue of the Americas. The owners sought to reimagine the building's lobby, public plaza and existing underground retail and commercial space, which was inaccessible from the street. An iconic architectural feature rebrands the property, replacing existing plaza fountains and serving both as a beacon for passersby and as a welcoming entry point to a newly activated underground space. The scheme sees a monumental elliptical glass canopy supported at an angle by an arching beam across the entrance. This lens-like oculus covers the similarly shaped sunken space, providing daylight, sky views and visual connection between the plaza above and the commercial space below.

住宅作品
RESIDENTIAL WORK

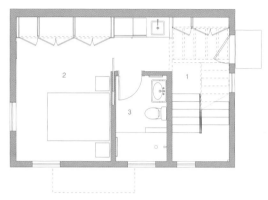
Ground Floor Plan

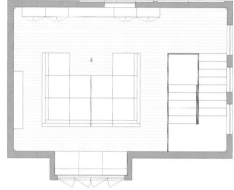
Second Floor Plan

1 Stair Hall
2 Guest Room
3 Bathroom
4 Media Room

贝德福德旅馆
Bedford Guesthouse

Bedford, NY
2010

Making new from old, this adaptive reuse project on an estate in the suburban New York village of Bedford converted a historic barn into a rustic-but-refined 74-square-meter guesthouse. The historic structure needed complete reconstruction, and the village required it to be rebuilt in its original form. To meet this challenge, the structure was painstakingly measured, then disassembled and rebuilt to carefully maintain its quirky slopes and mismatched corners. Finished with wood shingles, the barn will weather to its original silver-gray color. The two-story interior fulfills two functions, the upper level serving as a media room for weekend movie viewing, the lower containing guest entry, bedroom and bathroom. Inside, the mix of reclaimed materials includes a ceiling clad in wood repurposed from the original exterior siding.

1 Entry Hall / Gallery
2 Living Room
3 Dining Room
4 Kitchen
5 Master Suite
6 Bathroom
7 Guest Room

Floor Plan

中央公园西大道住宅
Central Park West Residence

New York, NY
2002

Designed to display the homeowner's important collection of 20th-century and contemporary photography, this two-bedroom, 190-square-meter apartment on Manhattan's Central Park West exhibits a distinct gallery-like atmosphere. In its public spaces, which comprise an entry hall and living and dining areas, creamy-gray poured-terrazzo floors and white walls combine with lighting specifically designed to highlight the artworks. In contrast, the private spaces, including the master suite, are finished in muted, natural tones. The bleached-wood floors, Venetian plaster, limestone, and French oak of these rooms create more personal, intimate environments.

Central Park West Residence

Central Park West Residence

Floor Plan

1 Entry
2 Living Room
3 Kitchen / Dining
4 Bedroom
5 Plunge Bath
6 Shower
7 WC

0 — 4.8m

科特兰庄园住宅
Cortlandt Manor Residence

Cortlandt Manor, NY
1995

Created for a contemporary-art collector, this 200-square-meter, one-bedroom home north of Manhattan serves as a personal retreat from the city. Its concept centers on four identical cubes—defined by sets of parallel brick walls—rotated in relation to each other as they line up to form a rectangular residence. Equal in size, each contains a single central function: eating, sleeping, bathing, or relaxing. Two-meter-wide glass enclosures connect the cubes, allowing square footage to expand in either direction—to create a larger living space, for instance. The house sits next to a wooded landscape, but the owner's need for wall space for art took precedence over large windows capitalizing on the views. A long, tree-lined drive does take advantage of the setting, however, shielding the structure from the public and making for a dramatic reveal.

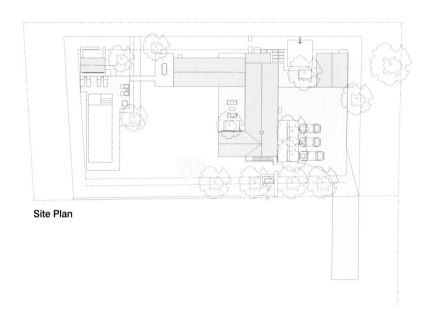

Site Plan

东汉普顿小镇度假屋
East Hampton Residence

Village of East Hampton, NY
2013

This 900-square-meter newly built house combines a traditional façade, inspired by its setting in the quaint village of East Hampton, with contemporary, light-filled interiors designed for 21st-century living. Formal, historic and private, the street-facing elevation displays the elegant proportions, and spare detail of Shaker architecture, while the rear proves more informal, modern, and open, its wide sliding-glass doors leading to a sun-filled backyard. Inside, natural materials and large windows visually connect this family country home to the landscape. The ground-floor great room's exposed wood rafters, open plan, and natural light—much of it from a 14-meter-long gabled skylight—continue the relaxed feeling. Upstairs, four bedrooms sit off a single hallway running the length of the house, allowing for easy transfer of light and air.

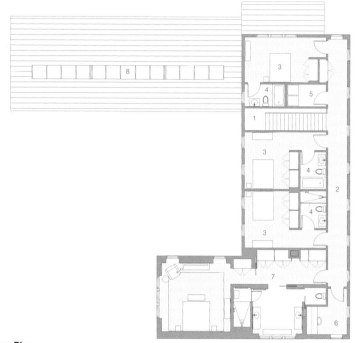

Second Floor Plan

1. Stair Hall
2. Hall
3. Bedroom
4. Bathroom
5. Laundry
6. Study
7. Master Suite
8. Skylight

Ground Floor Plan

1. Entry Hall
2. Living Room
3. Family Room
4. Kitchen
5. Pantry
6. Mud Room
7. Powder Room
8. Laundry
9. Bedroom
10. Bathroom
11. Terrace

East Hampton Residence

East Hampton Residence   209

Floor Plan

1 Living Area
2 Kitchen / Dining Area
3 Bedroom
4 Bathroom
5 Walk-in Closet
6 Dressing Area
7 Media Room

0   4.8m

格林威治村阁楼
Greenwich Village Loft

New York, NY
2009

A study in materiality, with texture and pattern explored in virtually every surface, this project in Manhattan's Greenwich Village transformed a raw commercial space into a 160-square-meter loft comprising a master suite, a convertible media-guestroom, and a great room with an open kitchen designed for entertaining. Clad largely in stainless steel and European walnut, the kitchen centers on a French marble–topped island whose angled, mirror-polished sides reflect a fumed-oak floor and clear acrylic stools. The eclectic living area features a herringbone-patterned, warmly colored silk rug, and custom crushed-velvet sofa, and lacquered credenza. In the bedroom, an eel-skin leather headboard sits atop a celadon silk carpet, while the media-guestroom includes a mid-20th-century Italian glass pendant light and raked-plaster walls. Calacatta marble slabs and floating glass walls define the bathrooms.

Greenwich Village Loft 217

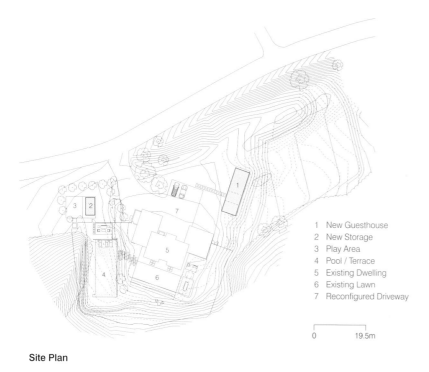

1 New Guesthouse
2 New Storage
3 Play Area
4 Pool / Terrace
5 Existing Dwelling
6 Existing Lawn
7 Reconfigured Driveway

0   19.5m

Site Plan

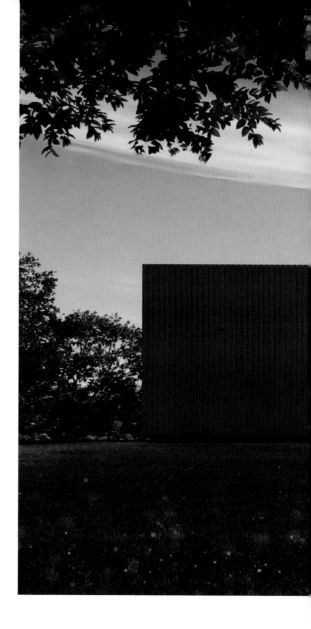

哈德逊旅馆
Hudson Guesthouse

Hudson, NY
2015

The master plan for this four-acre hillside site outside of the upstate New York town of Hudson includes a new guesthouse and pool adjacent to an existing contemporary home. The 6-by-14-meter pool takes advantage of the property's Hudson Valley views, while the guesthouse, nearly surrounded by a new meadow, forms an entry court with the main structure. A covered breezeway divides the guesthouse into two sides: one side a gym, the other a living-and-sleeping area for guests. The opening acts as a bridge between the sides, allowing for privacy as well as connection to the surrounding landscape. Clad in vertical wooden slats, the structure's simple construction, including exposed rafters and concrete flooring, features an elegant glass wall that maximizes the building's transparency and views.

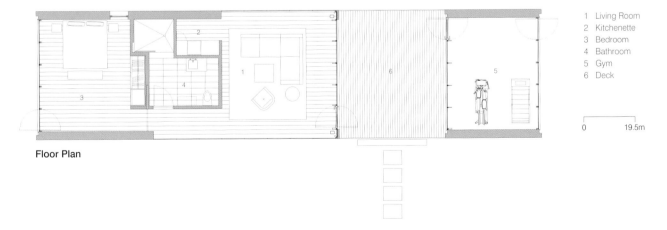

Floor Plan

1 Living Room
2 Kitchenette
3 Bedroom
4 Bathroom
5 Gym
6 Deck

0   19.5m

Hudson Guesthouse   221

1	Foyer
2	Kitchen
3	Living Room
4	Library
5	Study
6	Bedroom
7	Gallery
8	Bathroom
9	Walk-in Closet
10	Terrace

Floor Plan

伦敦阳台阁楼
London Terrace Penthouse

New York, NY
2011

Perched atop majestic London Terrace, a Depression-era development in Manhattan's Chelsea neighborhood, this 240-square-meter residence started out as a dark and claustrophobic four-bedroom apartment, emerging after a massive renovation as a spacious and light-filled one-bedroom penthouse. A simplified and largely open plan organizes the home on two axes, allowing views through its rooms and to skyline vistas beyond. A series of 22 new glass-and-steel French doors and windows replaces original small window openings, maximizing these views and allowing access to a wraparound terrace. Newly created spaces include a master suite with a bedroom and two bathrooms; an enfilade of rooms connecting the living room, library, and study; and an eat-in kitchen designed for easy entertaining. The interiors feature the clients' collection of antiques and art, as well as new and custom-created pieces and luxurious textiles.

曼哈顿连体别墅
Manhattan Townhouse

New York, NY
2014

The sweeping scope of this seven-story townhouse project on Manhattan's Upper East Side comprises six bedrooms; living, dining, and kitchen areas; a gym; and a subterranean indoor swimming pool, as well as a backyard and roof terrace. The homeowners purchased the historic building after its renovation and then commissioned custom interiors. Existing Italian-oak paneling set the stage for a selection of modern furniture carefully placed throughout the 840-square-meter house. Sheer draperies and muted fabrics bring a sense of calm quietude and serenity to the interiors, as does the spare quality of the overall scheme, which allows the clients' art collection—including works by Andy Warhol and Thomas Struth—to take center stage.

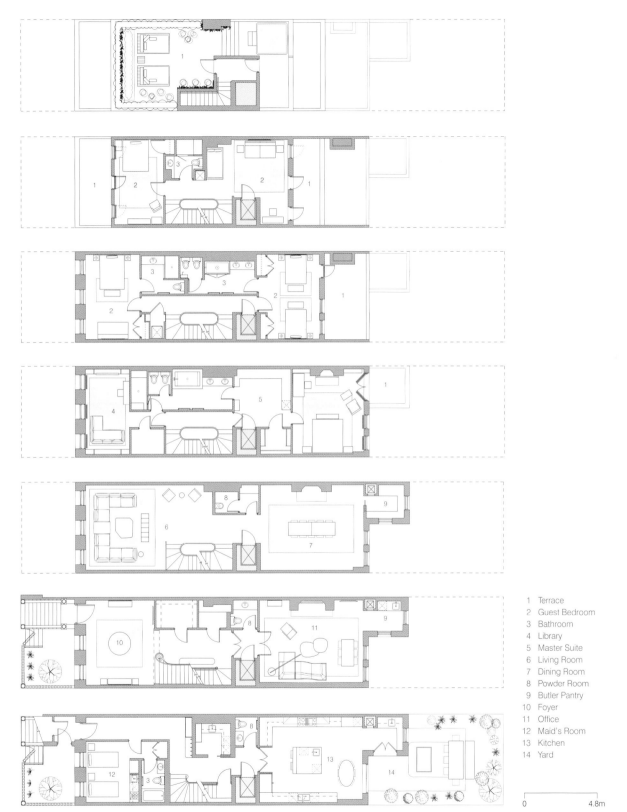

Floor Plans

1 Terrace
2 Guest Bedroom
3 Bathroom
4 Library
5 Master Suite
6 Living Room
7 Dining Room
8 Powder Room
9 Butler Pantry
10 Foyer
11 Office
12 Maid's Room
13 Kitchen
14 Yard

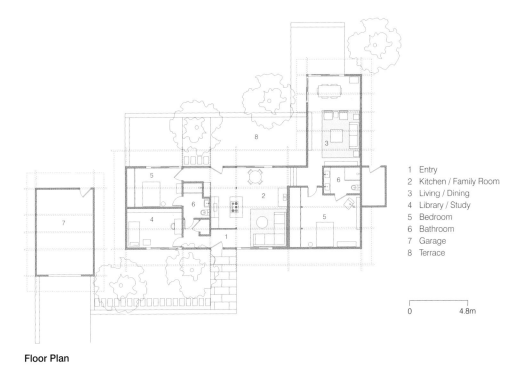

1	Entry
2	Kitchen / Family Room
3	Living / Dining
4	Library / Study
5	Bedroom
6	Bathroom
7	Garage
8	Terrace

Floor Plan

中世纪住宅
Mid Century Residence

Dutchess County, NY
2014

A sensitive restoration of a classic mid-20th-century residence in New York's Hudson Valley highlights and heightens the strong modernist lines of the 1968 home, while simultaneously adapting the two-bedroom, 610-square-meter house for contemporary living. Carefully selected furnishings comprise an era-appropriate collection of vintage pieces, including classic and rare designs by the likes of Greta Magnusson Grossman, Milo Baughman, Jens Risom, and Paul McCobb, all complemented by contemporary lighting and new custom creations. A redesign of the 7-acre property's grounds—which now prominently feature a wildflower meadow and a pool lined with slabs of native New York bluestone—completes the project.

Mid Century Residence

Mid Century Residence 247

Site Plan
0　　19.5m

红钩住宅
Red Hook Residence

Red Hook, NY
2015

This 465-square-meter residence lies at the end of a long, winding drive in the Hudson Valley town of Red Hook, NY, at the edge of a meadow and surrounded by woodlands. Conceived as an intimate private residence, able to both accommodate family gatherings and entertain guests, the house is organized along a series of interconnected perpendicular axes that create sightlines to carry the eye through the interior and out into the landscape beyond. These axes also divide the home into several pavilions, each with a separate function: there's a grand entertaining pavilion, master and guest-suite pavilions, a library pavilion that converts into a second guest suite, and a garage with a loft-like, second-floor space that can be used as an additional guest apartment.

Elevation

Section

Floor Plan

1. Entry
2. Living Room
3. Media Room
4. Dining Room
5. Kitchen
6. Pantry
7. Mud Room
8. Master Suite
9. Guest Suite
10. Study / Guest Room
11. Library
12. Garage
13. Up to Loft

Red Hook Residence

致谢
Acknowledgements

We would like to thank the outstanding members of our staff, without whom the projects we have designed would not be what they are, especially our associates, Takaaki Kawabata, who has been with the firm since 1998; Matthew Jasion, since 2003; and Camaal Benoit, since 2006.

Our appreciation also goes to the individuals whose careful attention to detail contributed greatly to the creation of this monograph, in particular our special projects director, Heidi Engstrom, who led the effort; to Young Ha Mok, whose beautiful drawings predominately illustrate the projects, along with those of Chie Ikeda and Angie Winston; and to Nicolas Michael / ArX NY for their beautiful renderings. We also extend our gratitude to all of the talented photographers who have captured our designs over the past 20 years, especially Scott Frances, Mikiko Kikuyama, Eric Laignel, Paul Warchol, and Michael Weschler.

Thanks, too, to Susan Becher, for her tireless support; to Paul Latham, our intrepid publisher; to Andrew Sessa, for his thoughtful introduction, which captures our intentions with succinct clarity; and to Rod Gilbert, for his elegant book design.

Lastly, we would like to acknowledge our inspiring clients, many of whom have been on this journey with us for much of the past 20 years, and all of whom have shared our passion for design and the role that it plays in our lives.

Mark Janson
Hal Goldstein
Steven Scuro

Partners, Janson Goldstein LLP

New York, NY – September 2015

图片版权
Photography Credits

Ben Rahn: 76–77, 79–85

Bill Waldorf: 136–139

Calvin Klein Inc.: 36–39

Dinex Group: 46–51

Eric Laignel: 74–75, 101–102

Grazia Casa: 193

Janson Goldstein LLP: 45, 256

Michael Desjardins: 74

Michael Weschler: 26–27, 29, 140–141, 143

Michelle Litvin: 106–111

Mikiko Kikuyama: 13–25, 31–35, 40–41, 62–73, 87–101, 103, 112–115, 128–129, 131–135, 158–163, 186–191, 212–219, 222–234, 236–247

Nikolas Koenig: 196–197

Paul Warchol: 55–57, 144–145, 165, 194–196

Rick Lew: 166–167

Rosky & Associates Inc.: 146–149

Scott Frances: 175–179, 201, 203–211

Toshi Yoshimi: 105

Every effort has been made to trace the original source of copyright material contained in this book. The publishers would be pleased to hear from copyright holders to rectify any errors or omissions.

The information and illustrations in this publication have been prepared and supplied by the architect. While all reasonable efforts have been made to ensure accuracy, the publishers do not, under any circumstances, accept responsibility for errors, omissions and representations express or implied.

Varick Street, 1995

Janson Goldstein LLP
Founded, 180 Varick Street, New York City, 1995
jansongoldstein.com